FLORA OF TROPICAL EAST AFRICA

COCHLOSPERMACEAE

B. Verdcourt

Trees, shrubs or rhizomatous herbs or subshrubs, often deciduous, with coloured juice and sometimes containing gum. Leaves alternate, palmatilobed, the lobes toothed or ± entire, often with domatia in the axils of the main nerves; stipules present, caducous. Flowers ± regular (very slightly zygomorphic), showy, hermaphrodite, paniculate or racemose, mostly golden-yellow. Sepals 4–5, free, imbricate, deciduous. Petals 4–5, free, alternating with the sepals, imbricate or contorted. Stamens numerous, the filaments free, equal or some longer than others; anthers 2-thecous, basifixed, linear, opening by introrse short often confluent pore-like slits. Ovary 3-, 4- or 5-carpellate, 1-locular with plate-like structures intruding into the locule, which is ramified with 3, 4 or 5 lobes corresponding to the number of carpels so that apical and basal sections appear axile and central sections appear parietal*; ovules numerous, anatropous; style simple, with small stigma. Capsule (2–)3–5-valved, the valves of the membranous endocarp separating from and alternating with those of the epicarp. Seeds glabrous or covered with woolly hairs, straight or mostly cochleate-reniform, with copious oil-rich endosperm; embryo large, conforming to the shape of the seed, with broad cotyledons.

A family of 2 genera and about 25 (but sometimes estimated as high as 40) species in the tropics and southern United States. Following Hutchinson (G.F.P. 2: 233 (1967)) the genus *Sphaerosepalum* Bak. has been excluded from the family.

COCHLOSPERMUM

Kunth, Malv.: 6 (1822); van Steenis in Flora Malesiana, ser. 1, 4: 61 (1949); G.F.P. 2: 233 (1967); Keating in Ann. Missouri Bot. Gard. 59: 282–296 (1973), *nom. conserv.*

Trees, shrubs or rhizomatous herbs or subshrubs. Sepals unequal. Filaments slender, equal or ± unequal. Capsule obovoid or ellipsoid, 3–5-valved, ± hairy or glabrous. Seeds cochleate-reniform, covered with long woolly hairs; embryo curved.

A small but widespread genus of about 15 species (*fide* van Steenis; 38 according to Hutchinson) in tropical and subtropical America with some in tropical Africa, SE. Asia, Malaysia and N. Australia. *C. religiosum* (L.) Alston (*C. gossypium* (L.) DC.) has been grown in East Africa (Kenya, Kilifi/Kwale Districts, Mazeras Nursery, Apr. 1930, *R. M. Graham* in *F.D.* 2334!) and has large palmatilobed leaves with broad acuminate lobes, woolly-white beneath and golden-yellow flowers.

C. tinctorium *A. Rich.* in Tent. Fl. Seneg.: 99, t. 21 (1831); Oliv., F.T.A. 1: 113 (1868), pro parte; Pilger in E. & P. Pf., ed. 2, 21: 317, fig. 141 (1925); F.P.S. 1: 155, fig. 92, pro parte (1950); F.W.T.A., ed. 2, 1: 185, fig. 70/A-G (1954); Hutch., Fam. Fl. Pl., ed. 3, fig. 77 (1973). Type: Senegal, Cayor, between Niaral and N'Denout, N'boro, *Leprieur* (P, holo.)

---

* Descriptions of the ovary as perfectly 3–5-locular are apparently erroneous.

1

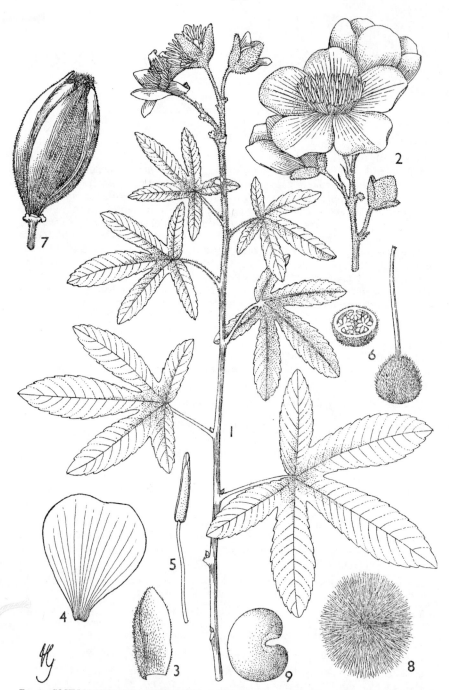

Fig. 1. *COCHLOSPERMUM TINCTORIUM*—**1,** flowering stem, × ⅔; **2,** inflorescence, × ⅔; **3,** sepal, × 2; **4,** petal, × 1; **5,** stamen, × 4; **6,** gynoecium. × 3; **7.** capsule, × ⅔; **8,** seed, × 2; **9,** seed, with hairs removed, × 4. 1, from *Chorley* 2050 & *Eggeling* 1228; 2–6, from *Purseglove* 1115; 7, from *Eggeling* 1228; 8, 9, from *Schweinfurth*. Drawn by Miss V. Goaman.

Rhizomatous suffrutex, with precocious flowers appearing on leafless shoots 5–15 cm. tall, the subsequent leafy shoots 30–90 cm. tall; stems pubescent above, glabrous and ridged below. Leaf-blades 2–12 cm. long, 2–16 cm. wide, usually deeply 3–5-lobed, the lobes (linear to) narrowly lanceolate to oblong or elliptic-oblong, 1–9·5 cm. long, 0·3–4·5 cm. wide, acute, ± entire or serrate, upper surface pubescent when very young, later glabrous or with hairs on the nerves near the base, lower surface glabrescent to pubescent when mature; petiole 1–6 cm. long; stipules linear, 5 mm. long. Flowers showy, appearing usually close to ground level but sometimes a few at the apices of leafy shoots, about 7–9 cm. in diameter; pedicels 1·5–4 cm. long, grey-tomentose and pubescent. Sepals elliptic-oblong to rounded, 1·1–1·7 cm. long, 0·5–1 cm. wide, grey-tomentose outside. Petals golden-yellow, rounded, 2·5 cm. long, 2·2 cm. wide. Anthers 4–6·5 mm. long; filaments equal or ± unequal, 4–9 mm. long. Ovary subglobose, 2·5 mm. wide, woolly or glabrous in some West African specimens; style 1–1·5 cm. long. Capsule fusiform-ellipsoid or obovoid, 4·5–6 cm. long, 2·5 cm. wide, glabrescent or tomentose, ± ridged; inner thin shiny layer straw-coloured. Seeds ± reniform but with a deep often narrow sinus, longest dimension 5·5 mm., shorter dimension 4·5 mm., 2·3 mm. thick, the sinus 1–1·5 mm. deep, covered with cream-coloured cottony hairs, shiny under the easily removed hairy testa. Fig. 1.

UGANDA. W. Nile District: Boroli, Pakelle, *Eggeling* 867! & Madi, Laropi, Mar. 1941, *Purseglove* 1115!; Acholi District: Atiak, Tero R., Dec. 1937, *Chorley* 2050!
DISTR. **Ul**; Senegal to Nigeria, Cameroun and Sudan
HAB. Grassland after burns; 600–1110 m.

SYN. *C. niloticum* Oliv., F.T.A. 1: 113 (1868) & in Trans. Linn. Soc. 29: t. 7 (1872); F.W.T.A., ed. 1, 1: 158 (1927). Type: Uganda, Madi, *Grant* 692 (K, holo.!)

INDEX TO COCHLOSPERMACEAE